1 Why do we need energy?

Energy transfer

We all need **energy** to stay alive and to be **active**. We get our energy from the food we eat. Energy can be **transferred** from one thing to another. For example when you lift a book from the floor to the table you transfer energy from your muscles to the book. An easy way to find out if energy has been transferred is to look for something which has changed. Look at the pictures to see what changes when energy is transferred.

Energy is needed to *lift* things (to change the height).

Energy is needed to make things *hotter* (to change the temperature).

Energy is needed to *start something moving* or to make a moving object *move faster* (to change speed).

Energy is needed to change the shape of things.

A Work in a small group. Make a poster to show energy being transferred. Use pictures from old magazines. Give your poster a title and label the changes taking place.

B Display the poster and as a group make a short presentation to the rest of the class.

Apparatus

☐ marker pens ☐ felt–tip pens
☐ sugar paper ☐ old magazines
☐ scissors ☐ glue

Q1 Write down six examples of where you or your friends have needed energy today.

2 Measuring energy

Joules

Energy is measured in units called **joules**. The symbol for joule is **J**. A joule is a very small amount of energy. You use one joule of energy when you lift one **newton** through a distance of one metre. A newton is roughly the weight of an average apple.

▲ 1J of energy is transferred from the person to the 1N weight when it is lifted 1m.

Q1 Copy this table.

	Height of weight (m)	Weight (N)	Energy transferred (J)
A		1	1
B		1	2

Apparatus

☐ clamp and stand ☐ G-clamp
☐ pulley ☐ tape measure
☐ 1N and 2N weights ☐ string

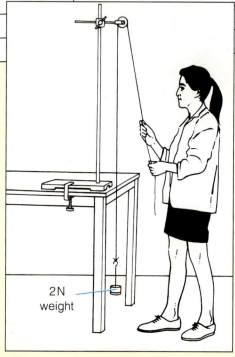

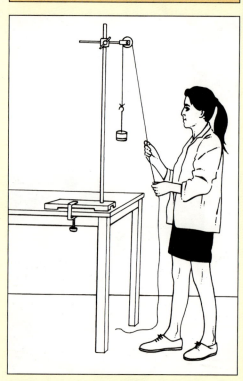

A Work in pairs. Set up the apparatus as shown. Transfer 1J of energy to the 1N weight. Write down the height of the weight in your table. ▲

B Transfer 2J of your energy to the 1N weight. Write down the height of the weight.

C Change the 1N weight for the 2N weight. Transfer 1J of your energy to the 2N weight. Write down the height of the weight. ▲

D Transfer 2J of your energy to the 2N weight. Write down the height of the weight.

E Transfer 4J of your energy to the 2N weight. Note the height of the weight. ▲

Q2 Copy this equation. Use the pattern in your table to fill in the blank spaces:

Energy transferred = ___ X ___

Q3 How much energy is needed to lift 4N through 1m?

Q4 How much energy is needed to lift 4N through 2m?

Q5 How much energy is transferred if the height moved is doubled?

Q6 How much energy is transferred if the weight lifted is doubled?

Q7 How much energy is transferred if the height and weight lifted are both doubled?

2 Measuring energy

Climbing the stairs

In this experiment you are going to measure the energy you transfer to your muscles to carry your weight up a flight of stairs.

Apparatus
- [ ] stairs [ ] bathroom scales
- [ ] metre rule

Q1 Copy this table.

Weight (N)	Number of steps	Height of one step (m)	Total height of stairs (m)	Energy transferred (J)

A Measure the **vertical** height of each step in metres. ▲

B Count the total number of steps in one flight of stairs. Work out the total height of the flight of stairs. ▲

C Measure your weight in newtons (1 kilogram = 10 newtons). Start at the bottom of the stairs and walk upwards. ▼

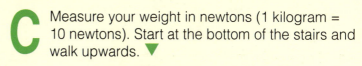

D Use your equation from **Q2** on page 2 to work out the total energy you transferred to your muscles as you climbed the stairs. ▲

Q2 How many flights of stairs have you climbed to get to your science lesson?

Q3 Work out the total energy you transferred to your muscles to get to your science lesson.

Q4 Do you use more energy when you run upstairs? Explain your answer.

3

2 Measuring energy

Energy stored in food

⚠ Ask your teacher to blow some custard powder into the Bunsen flame. When it burns the **molecules** of the food combine with oxygen and transfer their energy to their surroundings.

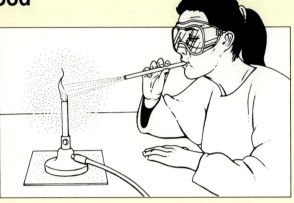

Apparatus

☐ peanut ☐ mounted needle
☐ test tube ☐ water
☐ measuring cylinder
☐ matches ☐ eye protection
☐ Bunsen burner ☐ heatproof mat
☐ 50 cm³ measuring cylinder

 Wear eye protection, and keep well away from the Bunsen flame when your teacher is blowing custard powder.

In this experiment you are going to transfer the energy stored in a peanut to some water.

Q1 Copy this table.

Temperature at start (°C)	Temperature at end (°C)	Temperature rise (°C)

A Using the measuring cylinder pour 25 cm³ of water into the test tube. Clamp the test tube and the thermometer as shown. Record the temperature of the water at the start. ▼

B Carefully place a peanut on the mounted needle. Do not point the needle at your fingers. ▼

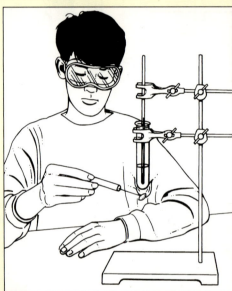

C Light the peanut using a Bunsen flame. When the peanut starts burning, hold it under the test tube straight away. ▲

D When it stops burning, record the temperature of the water straight away.

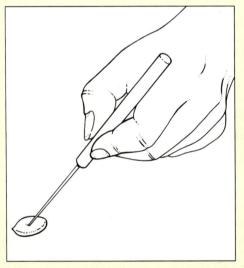

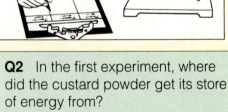

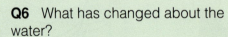

Q2 In the first experiment, where did the custard powder get its store of energy from?

Q3 What happens to stored energy in custard when we eat it?

Q4 In the second experiment, how much did the water temperature rise by?

Q5 Why do you have to do **C** and **D** straight away?

Q6 What has changed about the water?

Q7 What has been transferred to the water?

Q8 What is the original source of the energy stored in the peanut?

Q9 Has all the energy from the peanut been transferred to the water? (*Clue:* did anything other than the water get warm?)

Q10 What would happen to the rise in temperature if you made the volume of water larger? Explain your answer.

Q11 What would happen to the rise in temperature if you used a bigger peanut? Explain your answer.

2 Measuring energy

Energy for your body

Your body needs energy to keep you warm, to keep your heart beating, your lungs breathing and to make your muscles work. Your body also needs energy to make you grow. If the food you eat gives you more energy than your body needs you will put on weight.

▲ This person eats a lot, but gets a lot of exercise.

▲ This person eats a lot, but does not get much exercise.

▲ This person does not get much exercise, but does not eat much either.

Different people need different amounts of food, even if they get the same amount of exercise.

▲ Some people's bodies use more energy than others.

▲ A big person has more 'body' to keep warm, and more blood to pump around than a small person.

Q1 Does your body use energy when you are asleep?

Q2 What will happen if the food you eat gives you more energy than you need?

Q3 What will happen if the food you eat does not give you enough energy?

Q4 Give at least two examples of why different people need different amounts of energy.

2 Measuring energy

Apparatus
- ☐ felt-tip pens ☐ sugar paper
- ☐ snack wrappers
- ☐ food wrappers
- ☐ drink cans ☐ glue
- ☐ scissors

This bar chart shows how much energy an average person needs each day. Some people will need more energy than this. Some people will need less. (1000kJ = 1000000J)

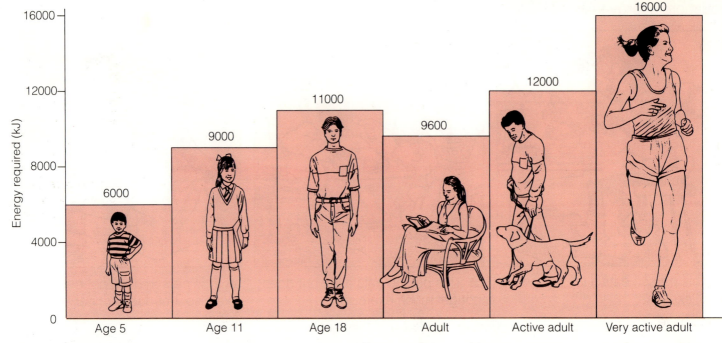

This chart shows you how much energy you get from different portions of food.

Cup of tea or coffee with milk (80kJ)	Chips (900kJ)	Bag of crisps (530kJ)
Teaspoon of sugar (80kJ)	Meat pie (1540kJ)	Apple (270kJ)
Cornflakes with milk (860kJ)	Mince and mashed potato (1160kJ)	Orange (170kJ)
Bread, butter and jam (700kJ)	Lasagne (1600kJ)	Twix (1100kJ)
Fried egg (500kJ)	Tuna fish (480kJ)	Kit Kat (1120kJ)
Sausage (880kJ)	Salad (100kJ)	Mars bar (1110kJ)
Bacon (1 rasher) (270kJ)	Salad cream (230kJ)	Tin of coke (700kJ)
Hamburger (in bun)(1570kJ)	Orange juice (300kJ)	Chocolate eclair (600kJ)
Pizza slice (910kJ)	Glass of milk (800kJ)	Rice pudding (750kJ)
Baked beans (470kJ)	Yogurt (750kJ)	
Peas (190kJ)	Apple pie and custard (1300kJ)	

Q1 Why does an 18 year old need more energy than an adult?

Q2 Use the chart to work out roughly how much energy you need each day.

Q3 Choose the foods you would normally eat in a whole day, including snacks and drinks. Add up how much energy you would get from this food.

Q4 Now write a new list of 'healthy' food that will give you the correct amount of energy for one day.

Q5 Use the food wrappers from crisps, chocolate bars, drinks or other snacks to make a poster to show how much energy you would get if you ate them. For example, you could show this sort of comparison in your poster.

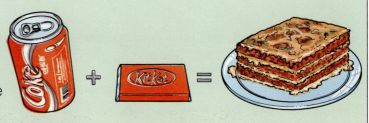

3 Energy sources

The Sun

Almost all the Earth's energy comes from the Sun. The Sun is a huge **nuclear reactor** which transfers a very large amount of energy to Earth and to the rest of the **Solar System**. The energy spreads out in all directions across space from the Sun's surface. The Earth is 'lucky' because it is at just the right place in our Solar System: it is neither too hot (like Mercury), or too cold (like Mars). This is one reason why Earth is the only planet in our Solar System which has life on it.

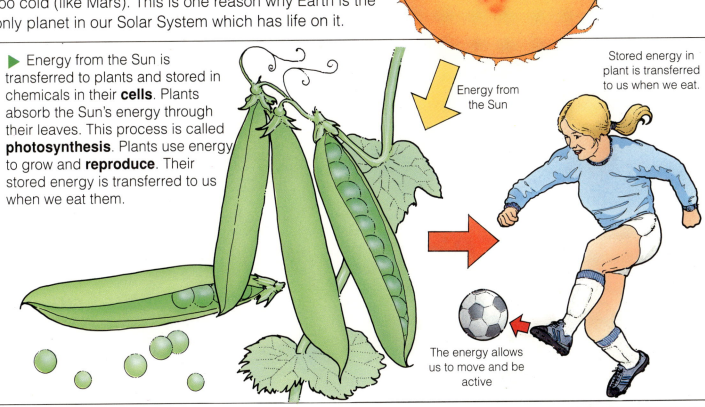

▶ Energy from the Sun is transferred to plants and stored in chemicals in their **cells**. Plants absorb the Sun's energy through their leaves. This process is called **photosynthesis**. Plants use energy to grow and **reproduce**. Their stored energy is transferred to us when we eat them.

Energy from the Sun

Stored energy in plant is transferred to us when we eat.

The energy allows us to move and be active

▼ Animals and humans are unable to take in the Sun's energy in the same way as plants. To get our energy we either:
- eat plants directly *or*
- eat animals who have eaten plants *or*
- eat animals who have eaten other animals who have eaten plants.

In this way all living things get their energy from the Sun.
In short: no Sun – no plants – no animals – no life.

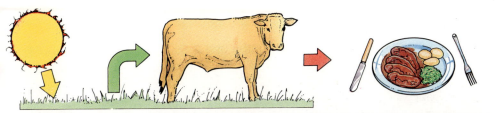

Q1 Where does the Earth get most of its energy from?

Q2 Why is the Earth the only planet with life on it in the Solar System?

Q3 a What part of a plant absorbs the Sun's energy?
b What is this process called?

Q4 Show a typical food chain for humans starting with the Sun.

Extension exercise 1 can be used now.

3 Energy sources

Fuels in everyday use

Coal, oil and gas are examples of **fuels** and so is food. These materials are called **sources of energy** because when they are burnt they combine with oxygen and *transfer* energy to their surroundings. We use this energy in many ways.

Q1 List all the sources of energy which you can see in the picture.

Q2 Which sources of energy do you use in your home?

Q3 In which months of the year do you use the most fuel in your home?

Q4 Describe the energy transfers you can see happening in the picture.

3 Energy sources

Non-renewable fuels

Coal, oil and gas are known as **fossil fuels** because they all come from fossilised plants or animals. Dead plants or animals are squashed by layers of rock. Over millions of years this pressure **concentrates** their stored energy into a smaller space. After **mining** and **drilling** for these fuels, the energy can be released by **burning**. The original source (the start) of all fossil fuels is the Sun. Without the Sun's light and heat there would be no plants, no food and therefore no animals. Coal, oil and gas are **non-renewable** fuels. This means that one day they will all run out. The picture below shows how coal and oil are formed.

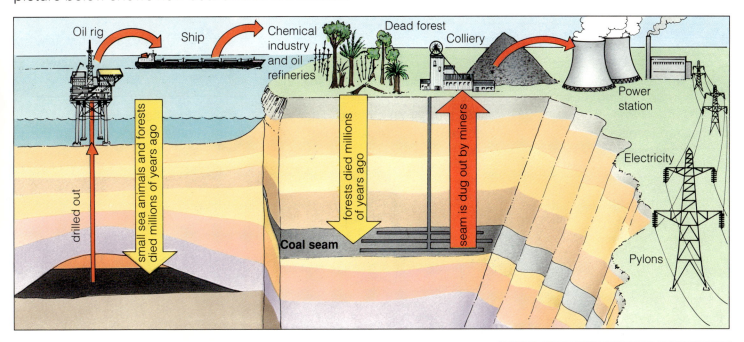

Coal
3000 million tonnes of coal are mined each year from the world's reserves. If this demand stays steady we have enough coal to last about 300 years. Figures show however that the demand keeps rising.

Gas
Gas supplies will start to run out around 2050. Over 65% of the world's gas reserves are in the Middle East and the Russian Federation.

Oil
If we keep using oil at the present rate it will run out in about 40 years. An alternative to oil for road and air transport will be hard to find. Oil is also the raw material for many different chemicals. Over 50% of the world's oil reserves are in the Middle East.

Uranium
30 000 tonnes of uranium are used each year. All the uranium will be used up by around 2030.

Q1 What is the original source of all fossil fuels?

Q2 What are fossil fuels made from?

Q3 List three fossil fuels.

Q4 How old will you be when oil runs out? How will this affect you?

Q5 Why will fossil fuels eventually run out?

Q6 Name a fuel which is non-renewable but not a fossil fuel.

Q7 Why will the Middle Eastern nations become more powerful as our fossil fuels run out?

3 Energy sources

Burning fuels

When things burn they combine with oxygen. The energy stored in them is transferred to the things around them, which get warmer. You are going to find out how much energy is transferred when different kinds of fuel burn.

Q1 Copy this table.

Fuel	Temperature of water at start (°C)	Temperature of water at end (°C)	Temperature change (°C)
Methylated spirits (burner X)			
Paraffin (burner Y)			
Candle			

A Fix a boiling tube above fuel burner X, using a clamp and stand. Burner X contains methylated spirits. ▼

B Pour 30 cm³ of water into the boiling tube. Take the temperature of the water and record it in the second column of your table. ▼

C Light the wick using a lighted splint, and start the stop clock. ▼

Apparatus

- ☐ boiling tube
- ☐ clamp and stand
- ☐ 50 cm³ measuring cylinder
- ☐ thermometer ☐ candle
- ☐ metal fuel burners X and Y with stoppers and wicks
- ☐ tin lid ☐ stop clock
- ☐ eye protection ☐ heatproof mat

 Wear eye protection.

 Methylated spirits is highly flammable. Take great care not to spill any near a naked flame.

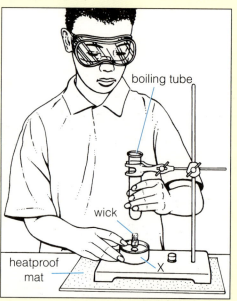

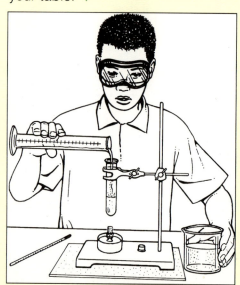

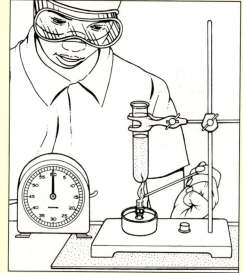

D When the fuel has burned for four minutes blow the flame out. Take the temperature of the water and record it in the third column of your table.

E Calculate the temperature difference and record it in the fourth column of your table. Pour away the water in the boiling tube.

F Repeat **A** to **E** with burner Y. Burner Y contains paraffin. ▶

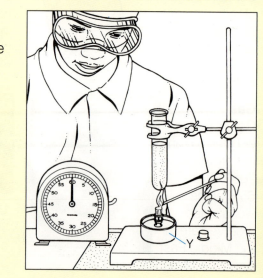

3 Energy sources

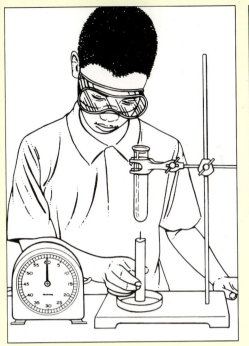

G Replace burner Y with a candle in a tin lid. ▲

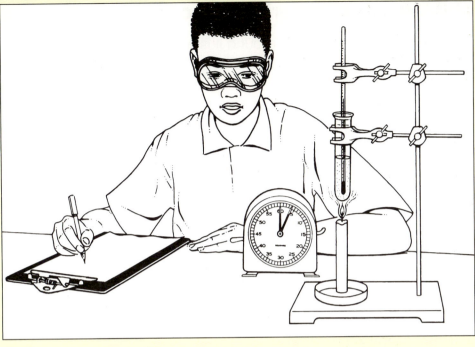

H Repeat **B** to **E**. ▲

Q1 Which fuel made the water in the boiling tube the hottest?

Q2 Write a list of the three fuels, starting with the one that made the water the hottest, and ending with the one with the coolest water.

Q3 What happened to the outside of the boiling tube when you heated it?

Q4 Why do you think people do not use paraffin for cooking at home?

Q5 Methylated spirit and paraffin are often used in camping stoves. Methylated spirit costs 110 pence per litre, and paraffin costs 180 pence per litre. Which fuel is the best value for money?

Q6 Was all of the energy stored in the fuel transferred to the water? (*Clue* did anything else get warm?)

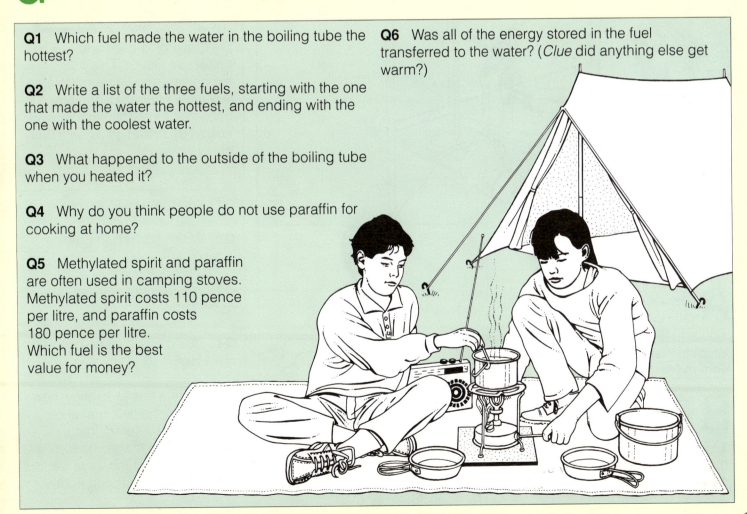

3 Energy sources

Making useful material from oil

When oil is drilled out of the ground it is a mixture of many different substances. It is of no use until the mixtures are separated. This is done by boiling the oil. The mixtures boil at different temperatures. The substance which boils at the lowest temperature boils off first and is collected. The substance with the highest boiling point boils off last. Separating a mixture by boiling is called **distilling**. The different substances are called **fractions**. Your teacher will show you how useful fractions are collected.

Fraction boils about (°C)	Uses of crude oil fractions
50	Calor gas
110	Petrol, pesticides, drugs, plastics, fertilisers, solvents, detergents
180	Paraffin, kerosene (jet fuel), white spirit
260	Diesel (truck and bus fuel), central heating oil
340	Lubricating oils and grease, candles, polishes
400	Tar for road surfaces, fuel for power stations, waterproof roofing materials (residues and bitumen)

Your teacher will explain how crude oil is distilled in an oil refinery.

Apparatus

- ☐ crude oil ☐ clamp and stand
- ☐ side-arm test tube
- ☐ test tube rack ☐ 4 test tubes
- ☐ test tube holder ☐ splints
- ☐ heatproof mat ☐ tin tray
- ☐ holed stopper with fitted thermometer ☐ protective gloves
- ☐ marker pens ☐ felt-tip pens
- ☐ paper ☐ sugar paper ☐ glue
- ☐ magazines ☐ eye protection

Wear eye protection and protective gloves. This experiment should be performed in a fume cupboard. The first fraction is highly flammable, and should be boiled off using a bath of hot water.

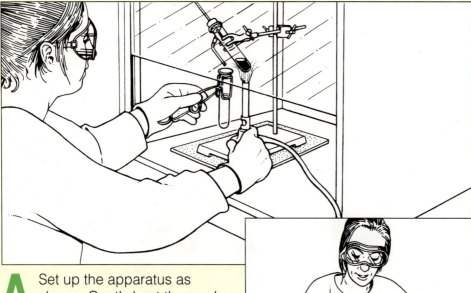

A Set up the apparatus as shown. Gently heat the crude oil. Hold a test tube under the side arm to collect the fractions, keeping your hands away from underneath the crude oil. ▲

B Collect the fractions between the following temperatures in four separate test tubes:
1. Up to 100°C
2. 100°C to 150°C
3. 150°C to 200°C
4. 200°C to 250°C.

C Pour the liquid in test tube 1 on to a tin lid or tray and note how easily it burns. Repeat this for liquids 2 to 4. ▲

Q1 What colour was the crude oil?

Q2 Which liquid burned most easily?

Q3 Which liquid smelled like petrol?

Q4 Could crude oil be used to run a car?

Q5 Make a poster to show how crude oil is distilled in a refinery, and how the different fractions are used. Use magazine pictures to show some of the ways we use the materials made from oil.

3 Energy sources

Solar energy

You are going to make a heater using energy from the Sun. We call energy from the Sun **solar energy**. Solar energy is **renewable energy** because we cannot use it up.

▶ Water is pumped through **solar panels** and is warmed up by the Sun. Solar panels are fixed to south facing roofs. The hot water can be used for washing or heating.

◀ **Greenhouses** make use of the Sun's energy. The Sun shines on to the greenhouse and warms it up. The glass walls and roof trap the heat inside, and keep the plants warm. Exotic flowers, fruits and vegetables that would normally be grown only in hot countries can be grown by **market gardeners** in Britain.

Q1 Copy this table.

Temperature of tap water (°C)	Temperature of water from tube (°C)	Temperature rise (°C)

Apparatus

☐ black tubing ☐ lamp
☐ thermometer ☐ beaker
☐ sticky tape ☐ strong card

A Pour some tap water into a beaker and measure its temperature. ▼

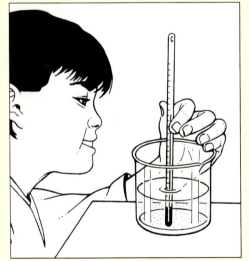

B Seal one end of the black rubber tubing with sticky tape. Fill the tube with water from the tap. Dry any water off and seal the other end with sticky tape. ▼

C Tape one end to the centre of a piece of card and make the tube into a coil. If it is a sunny day put the coil in direct sunlight, otherwise shine a lamp on it in the laboratory. ▼

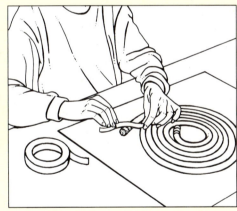

D At the end of the lesson pour the water out of the tube into a beaker and measure its temperature straight away.

E Design and carry out an experiment to show that on a hot day it is hotter inside a greenhouse than outside. Write a report showing your method and any measurements you took.

Q2 What happened to the temperature of the water in the tube?

Q3 If you left the tube in the Sun for a longer period of time what would happen to the water inside it?

Q4 If you had a bigger tube and more time give two examples of how you could use the water.

Q5 How do greenhouses help market gardeners?

Q6 Not many people use solar heaters in this country. Why do you think this is?

Q7 Why do you think solar energy will become important in the future?

Q8 Why is it that some countries can rely almost entirely on solar energy for use in their homes?

3 Energy sources

Renewable energy

Energy from rotting plants
▶ When plants die **bacteria** cause them to rot. Some of the Sun's energy stored in the plants is released as they rot and so the plant material warms up. If you help someone to dig out a compost heap you will find that it is warm in the middle.

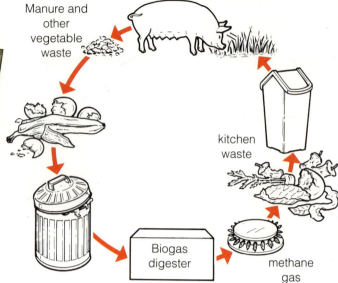

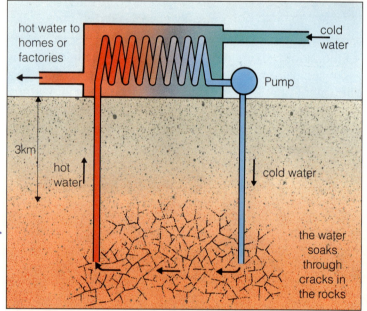

▲ **Energy from hot rocks**
Rocks underground are warmer than the rocks at the Earth's surface. In Cornwall this energy is used to heat water. This is done by drilling two holes and pumping cold water down one of them. Energy is transferred from the rocks to the water, which then warms up. The warm water comes up the other hole and is used to heat houses or factories. Energy obtained from hot rocks is called **geothermal energy**.

The Sun and the rocks in the Earth will not cool down for billions of years, and if we grow plants and keep animals, there will always be decomposing plants or manure to use as energy sources. We call energy we get from these sources **renewable energy** because it will not run out.

▲ **Methane** gas is released as plants rot. This can be collected and burned to give heat. Many villages in China use animal **manure** to produce gas in this way. The **decomposed** manure can be used as a fertiliser afterwards.

▼ In some parts of the world (for example India) cow pats are dried out and burnt as fuel.

Q1 What is geothermal energy?

Q2 Draw a flow chart to show how methane gas is produced from waste.

Q3 Can you think why there is still some energy left in the cow pats shown in the photograph?

Extension exercise 2 can be used now.

4 Storing and transferring energy

Apparatus

☐ spring ☐ connecting belt
☐ dynamo ☐ lamp and lamp holder
☐ heatproof mat ☐ Bunsen burner
☐ gauze ☐ tongs ☐ tripod
☐ steam generator ☐ voltmeter
☐ propeller ☐ eye protection

 Wear eye protection.

 Keep your fingers clear of the spring.

Dynamos

You are going to find out about some different ways of storing and transferring energy. A spring can be used to store energy. Stored energy is called **potential energy**. When the spring unwinds, its potential energy is changed to kinetic energy (movement).

A **dynamo generates** (makes) electricity when it is turned. The energy used to turn the dynamo is transferred to the lamp. When energy is transferred in this way we call it **electricity**. The lamp transfers the energy to its surroundings by warming and lighting them.

A Carefully wind up the spring. Lock it with the ratchet arm to stop it unwinding. Keep your fingers clear of the spring. Unlock it and watch what happens to the **axle** and wheel connected to it. ▼

B Connect the spring to the dynamo using the connecting belt. Connect a lamp to the dynamo. Repeat **A**. Watch what happens to the lamp. ▼

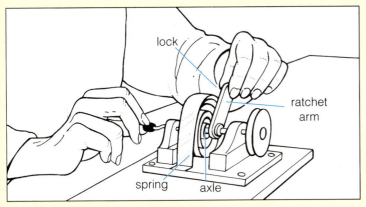

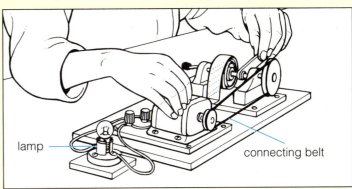

C Your teacher will show you how electricity is generated from steam. ▼

Q1 What did you transfer to the spring when you wound it up?

Q2 What did the spring make the axle and wheel do as it was unwinding?

Q3 What part of the system transfers the energy from the spring to the dynamo?

Q4 What does the dynamo transfer its energy to?

Q5 What do we call energy transferred in this way?

Q6 What does the lamp produce with the energy?

Q7 Where did the energy come from to make the steam?

Q8 What did the steam do to the propeller?

Q9 Draw a flow chart to show how the energy is transferred in **C**.

4 Storing and transferring energy

Motors

In this experiment you are going to use a motor to lift a weight.

Q1 Copy this table.

Weight (newton)	5N
Distance moved by the weight (metre)	
Time taken to lift the weight (second)	
Ammeter reading (ampere)	
Voltmeter reading (volt)	
Total energy transferred to motor (joule)	
Total energy transferred to weight (joule)	

Apparatus
- ☐ motor ☐ power supply
- ☐ weights and holder ☐ string
- ☐ pulley ☐ leads ☐ 2 G-clamps
- ☐ lamp and lamp holder
- ☐ wire hook ☐ ammeter
- ☐ voltmeter ☐ metre rule
- ☐ stop clock

A Set up your apparatus like this and ask your teacher to check it for you. ▼

B Switch on the motor and start your stop clock. Your partner must read the ammeter and voltmeter while the motor is running. Stop the clock when the weight has been raised one metre.

C Keep the weight in its high position by using the wire hook. Disconnect the power supply.

D Connect the lamp to the dynamo. (When you turn a motor without an electrical power supply it becomes a generator, (a dynamo). Undo the wire hook and let the weight fall. ▼

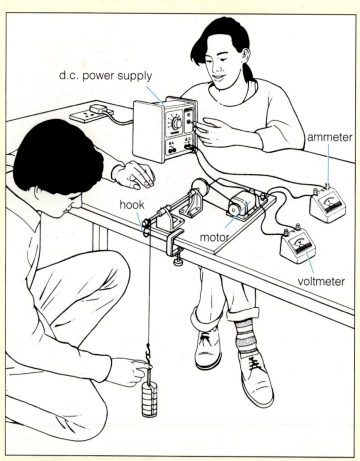

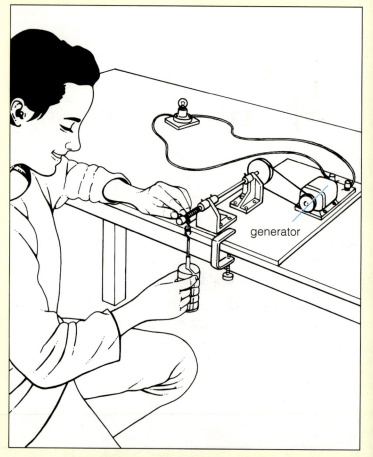

Q2 What happens to the lamp when the weight falls?

Q3 Where has the energy come from to drive the generator?

Q4 Has all the energy of the falling weight been transferred to the lamp? Give a reason for your answer.

Q5 Disconnect the connecting belt and watch the weight fall. Can you explain why it falls faster than in **D**?

4 Storing and transferring energy

How much energy did the motor use?

Electricity transfers energy to the motor which then transfers energy to the weight by lifting it. When the weight is at the top position it has its own supply of stored energy. This stored energy is called **gravitational potential energy**. It has many uses.

▲ Potential energy can be converted to **kinetic energy** (movement). The potential energy stored in this ball is converted to kinetic energy as it falls. It is going to knock this building down.

▲ The gravitational potential energy stored in this falling roller coaster is converted to kinetic energy and is enjoyed by the riders.

To find out how much energy was transferred to the motor when it was lifting the weight you need to use the following formula. Record the result of your calculation in your table from page 16.

Total energy transferred to the motor = ammeter reading X voltmeter reading X time

To find out how much potential energy is stored in the weight you need to use the following calculation. Record the result of your calculation in your table.

Energy transferred to the weight = weight X distance moved

Q1 Which has the largest energy value: energy transferred to the motor or energy transferred to the weight?

Q2 What do you think has happened to the energy which was not transferred to the motor?

Q3 Write down three examples of where potential energy is converted to kinetic energy in everyday life.

Extension exercise 3 can be used now.

17

4 Storing and transferring energy

How efficient was the motor?

Conservation of Energy

Energy is **conserved**. This means that energy cannot be destroyed; it can only be transferred to different materials around it. Think of what happens when you heat a pan of stew. The **useful** energy heats the stew but the pan, the cooker and the air around it get hot as well. We call this **wasted** energy because it does not do a useful job. **Efficiency** measures compare the total energy transferred with the 'useful' energy transferred.

Some of the energy used by the motor was transferred to gravitational potential energy in the raised weight. We say this was **useful energy** because it did a useful job (in lifting). The rest of the energy was used to turn the motor, wheel, axle and connecting belt, and to push the air out of the way as it lifted the weight. We call this 'wasted' energy. It is why a motor can never be 100% efficient. To calculate the efficiency of the motor you need to use the following equation. The numbers are in your table from page 16.

$$\text{Efficiency (\%)} = \frac{\text{Total energy transferred to weight}}{\text{Total energy transferred to motor}} \times 100\%$$

▼ This is an **energy transfer diagram** for the motor. It shows how much energy was usefully transferred and how much was wasted. The width of the arrow represents the amount of energy. Draw a diagram like this one for your motor. Ask your teacher to help you draw the width of the arrows to **scale**.

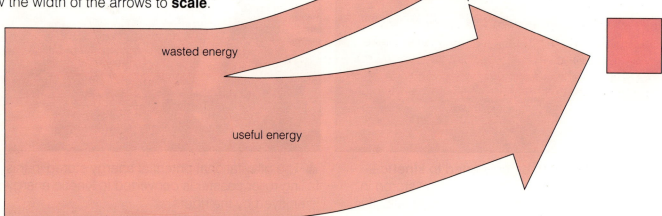

total energy into motor 100%

wasted energy

useful energy

Q1 What is the efficiency of the motor?

Q2 Why can a motor never be 100% efficient?

Q3 In one minute a lamp uses 6000 J of energy. Only 120 J of this energy are transferred to the light it emits.
a What has happened to the other 5880 J?
b Draw and label an energy transfer diagram for a lamp.
c How efficient is the lamp?

Q4 A car's engine is only about 15% efficient. Out of every 100 litres of the petrol put in the petrol tank only 15 are used to make the car move.
a What has happened to the energy stored in the other 85 litres?
b Draw an energy transfer diagram for a car.

Q5 What do you notice about the energy at the start and the total energy at the end in the situations in **Q3** and **Q4**?

Q6 We are told we cannot destroy energy so why do we worry about wasting it? What are we really wasting?

18

4 Storing and transferring energy

Generating electricity

Most of the electricity used in Great Britain comes from fossil-fuelled power stations. Most power stations burn coal to boil water and make steam. This steam is used to turn a **turbine**. The turbine turns the generator which makes electricity. The steam is cooled to turn it back into water. It goes back to the boiler to be heated again. The steam is cooled in the **condenser**.

Apparatus

☐ handout 1 ☐ scissors
☐ glue ☐ paper ☐ felt-tip pens

How a power station works

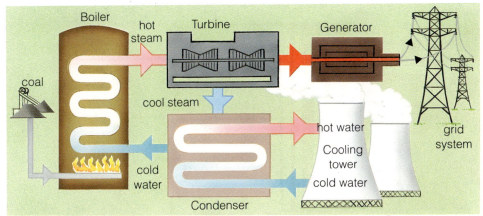

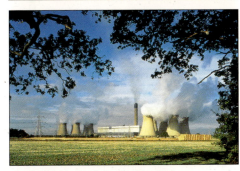

A fossil-fuelled power station

The model on page 15 is like a very small power station. The steam turns the propeller which turns the dynamo which makes electricity. Your teacher will give you handout 1.

Nuclear reactions can be used to make steam instead of fossil fuels. In a nuclear reaction **atoms** of uranium split up to make new elements. A lot of energy is transferred when this happens. Everything near the reaction gets hot. This heat is used to make steam in the power stations. Uranium is mined from the ground, just like coal. One day the uranium will run out.

1 Nucleus of Uranium 235 atom absorbs neutron

2 This makes it less stable

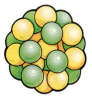
3 The nucleus splits

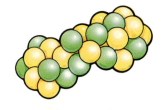
fission fragments

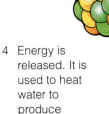
4 Energy is released. It is used to heat water to produce steam

5 Neutrons are also released which are used to continue the process

A nuclear power station

Nuclear reactions make the things around them **radioactive**. Radioactivity is dangerous, and making sure it does not harm people or animals is very expensive.

| Q1 Complete handout 1. | Q2 Is nuclear energy a renewable or non-renewable energy source? | Q3 At what stage in the nuclear reaction is energy released? |

4 Storing and transferring energy

Energy in? Energy out?

▼ This diagram shows the different sources of energy which are used to generate electricity in our power stations. It also shows the different ways we use electricity.

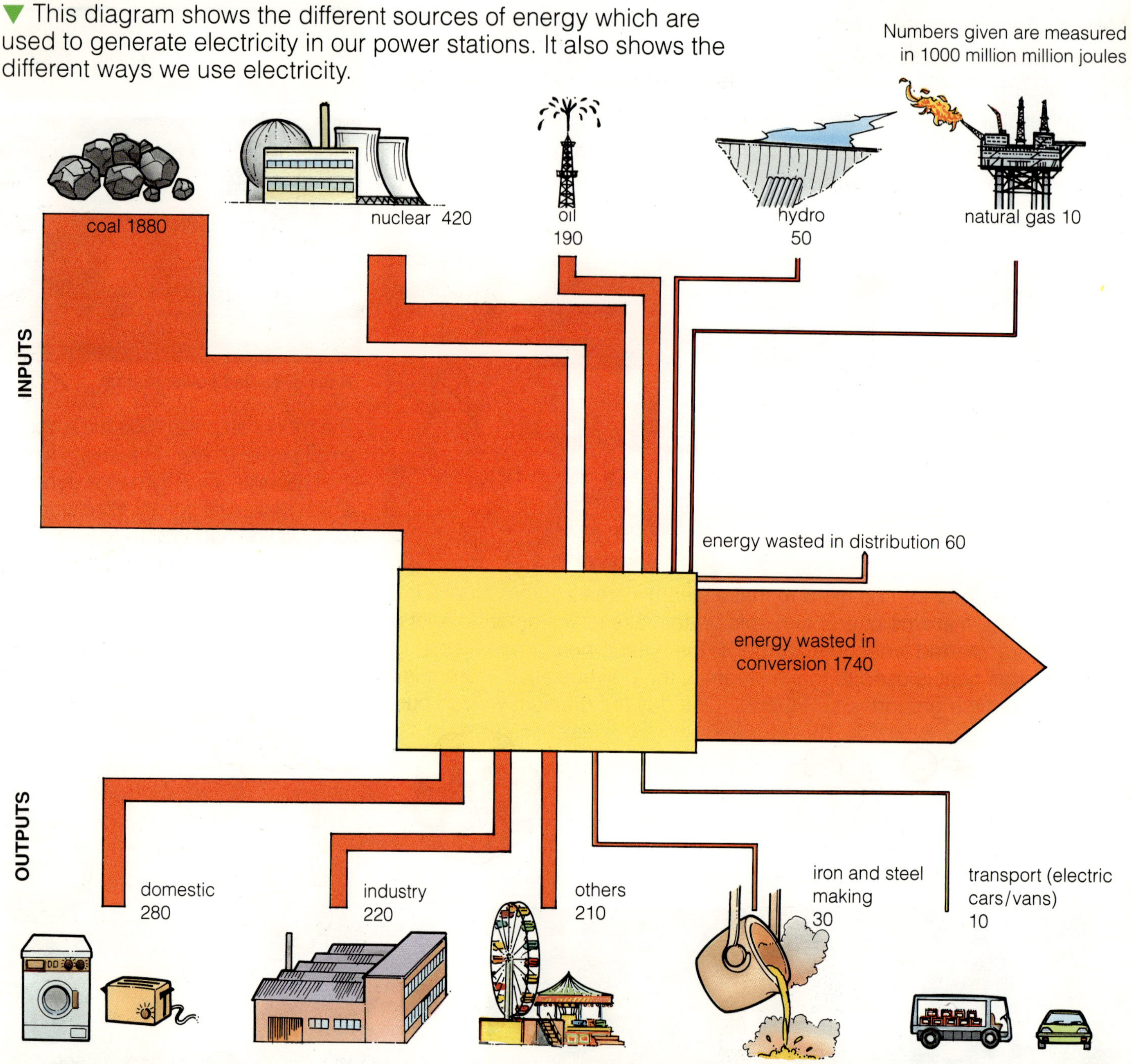

Numbers given are measured in 1000 million million joules

INPUTS

coal 1880
nuclear 420
oil 190
hydro 50
natural gas 10

energy wasted in distribution 60

energy wasted in conversion 1740

OUTPUTS

domestic 280
industry 220
others 210
iron and steel making 30
transport (electric cars/vans) 10

Q1 Draw a bar chart to show how much energy is transferred from each energy source.

Q2 Draw a pie chart to show the different uses of energy. Ask your teacher to help.

Q3 List the uses of electrical energy in your home and school.

Q4 What is the total input energy to power stations?

Q5 What is the useful output energy?

Q6 How much energy is wasted?

Q7 What do you think happens to the wasted energy?

Q8 Calculate the efficiency of the power stations.

Extension exercise 4 can be used now.

4 Storing and transferring energy

Electricity from renewable sources

▶ Wind

Windmills have been used for hundreds of years to grind corn and pump water. Energy in the wind can also be used to turn turbines to generate electricity. Modern windmills (turbines) are carefully designed to use as much of the energy in the wind as possible. It would take about 6000 turbines to replace one fossil-fuelled power station. A lot of land would be needed. Wind turbines would have to be placed in windy places like mountain tops or wide open spaces. They are very big and noisy.

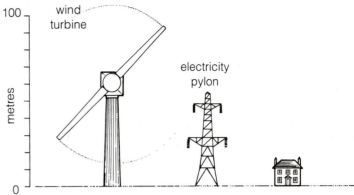

Wave

Waves are caused by winds blowing across the sea. The energy in waves can be captured by using large **floats** which move up and down with the waves. This movement is used to generate electricity.

◀ Hydroelectric

People who lived near rivers and streams would use waterwheels to drive machinery, for grinding corn or for making things. Energy in falling water can be used to generate electricity. This is called **hydroelectricity**. Hydroelectric power stations are built in high places, up in mountains or hills.

In this experiment you are going to make a wind generator.

A Connect a propeller and lamp to the dynamo. Use the blower to turn the propeller. Watch the lamp and listen to the propeller. Notice what happens if you change the wind direction. ▼

Apparatus

- ☐ propeller ☐ dynamo
- ☐ lamp and lamp holder
- ☐ leads ☐ blower

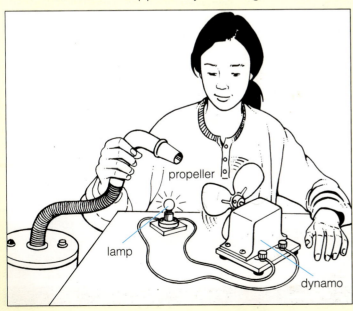

Q1 What happened to the lamp?

Q2 Why do you think people might not want wind turbines near their homes?

Q3 What happened when you changed the wind direction?

Q4 Is wind a reliable source of energy?

Q5 Can you think of any other things that use wind energy? *Hint:* leisure activities.

Q6 Can you think of some reasons why people might not want a hydroelectric power station near their home?

4 Storing and transferring energy

Renewable energy sources

The energy transferred using wind turbines, hydroelectric generators or waves originally came from the Sun. The Sun's energy evaporates the water which then falls as rain and fills up the rivers and lakes. The Sun's energy heats up the air. This causes the wind to blow. Wind blowing across the sea makes waves.

In this activity you are going to learn about the sources of renewable energy.

Apparatus
- ☐ felt-tip pens ☐ paper
- ☐ card ☐ safety pins
- ☐ sticky tape ☐ scissors
- ☐ sugar paper ☐ marker pens

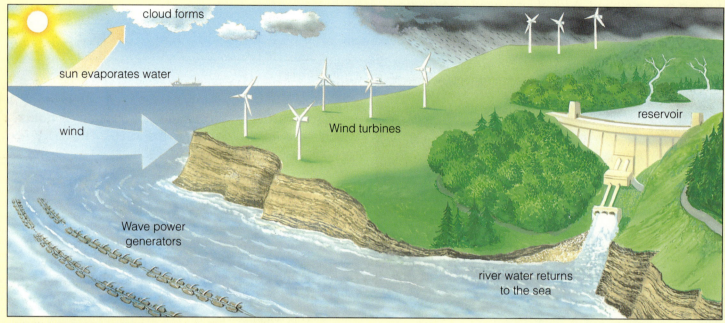

A Work with your friends. Discuss your ideas and make a poster to show how we can generate electricity without using up fossil fuels. ▼

B Design and make a badge showing one form of renewable energy. Wear your badge so the others can find out about renewable sources of energy.

Q1 List three sources of renewable energy which are used to generate electricity.

Q2 Where does this renewable energy originally come from?

Q3 In Britain which months of the year would be best for generating electricity from falling water, waves and wind?

Q4 Can you think of an **environmental** advantage which hydroelectric power has over wave or wind power?

Q5 Why could we not rely entirely on the three sources of energy you have just read about? Think of at least two reasons.

Extension exercise 5 can be used now.

4 Storing and transferring energy

Power stations

Electricity generated in power stations is sent around the country by a system of power lines called the **national grid**. All the big power stations supply energy to the national grid, and most people who use electricity take it from the national grid.

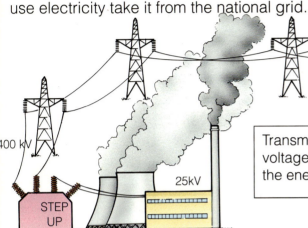

Transmitting electricity at high voltage is very efficient. 97% of the energy is transferred.

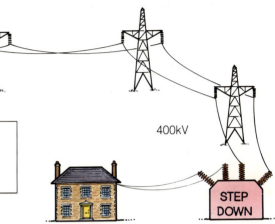

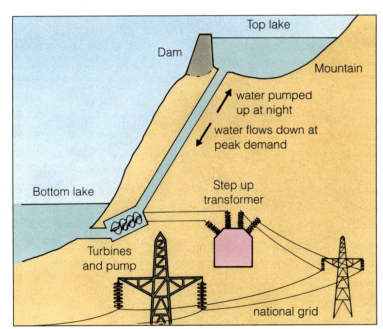

◀ Sometimes power stations cannot make enough electricity. When this happens they buy electricity from other power stations. Some power stations store energy at night. They are called **pumped storage power stations**. At night some of the energy transferred by the national grid is used to pump water back up to the top lake. This water will fall to the lower lake to make more electricity when people need it most (for early morning, tea time and breaks in TV programmes).

Q1 Look at the graph of demand for electricity during the football cup final. What time was the interval during the match?

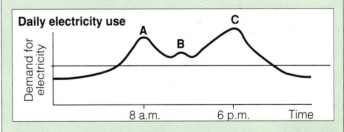

Q2 What do you think is happening at A, B, and C in this graph of daily electricity use?

Q3 What kind of energy does the water in the top reservoir of a pumped storage power station have?

Q4 What do power stations do if they cannot supply enough electricity?

▼ The people who control the power stations can tell when a very popular programme is on the television. This is because when the advertisements come on, or the programme finishes, millions of people make hot drinks and the demand for electricity suddenly increases. This graph shows the demand for electricity while the football cup final match was being shown.

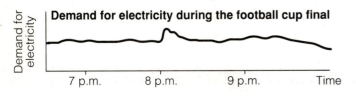

23

4 Storing and transferring energy

Electricity at home

When people say 'Save Energy' they really mean 'Save Fossil Fuel'. You are going to find out how the design of a kettle can help to save fossil fuels.

Apparatus

☐ a round kettle ☐ a jug kettle
☐ clean measuring jug
☐ stop clock

Electricity transfers energy to many devices which produce light, heat, sound and movement. The devices do many useful jobs. In some electrical machines the energy is used to do more than one thing, for example a stereo system.

A Fill the kettle until the element is just covered. Pour the water into the measuring jug to see how much there is, then pour it back into the kettle. Record the volume in your table. ▼

Q1 Copy these tables.

Minimum volume of water (cm³)	Time to boil (s)	Power rating (W)	Energy transferred (J)

B Switch the kettle on and use the stop clock to measure how long it takes to boil the water. Record the time in your table. ▼

C Repeat **A** and **B** using the jug kettle.

D Record the **power rating** of the two kettles in your table. It is the number measured in **watts (W)**, written on the side of the kettle by your teacher. The power rating tells you how much energy the kettle uses every second. ▶
Use this equation to work out the energy transferred:

Energy transferred = power × time
(W)   (s)

Record your result in your table.

Q2 Which kettle used the most energy?

Q3 How does the design of the jug kettle help to save non-renewable fuel?

Q4 If you only wanted to make one cup of tea, which kettle should you use?

Q5 Make a list of the electrical devices in your home, and say what the electricity is used to do.

4 Storing and transferring energy

Keeping warm

▶ A good **conductor** is a material which lets energy flow through it easily. Good conductors often feel cold when you touch them because they conduct energy away from your hands quickly. When energy is transferred quickly from our bodies to our surroundings we feel cold. In winter we try to stop energy leaving us by wearing clothes which are bad conductors of energy. We also wear lots of layers of clothing which trap layers of air. Trapped air is a bad conductor of energy. Bad conductors are called good **insulators**. Good insulators keep us warm.

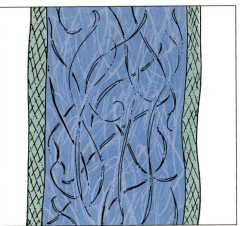

Ski jacket fibres trap lots of air.

In this experiment you are going to find out what happens when things feel 'hot' or 'cold'.

A Place your hands on the plastic bowl. Does it feel warm or cold? ▼

B Place your hands on the metal saucepan. Does it feel warm or cold? ▼

C Take your shoes off. Stand on the carpet tile. Does it feel warm or cold? ▶

D Stand on the quarry tile. Does it feel warm or cold?

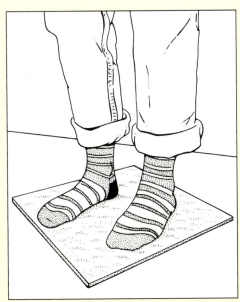

Apparatus

☐ large plastic bowl
☐ saucepan ☐ carpet tile
☐ large quarry tile

Q1 List the things in your investigation which felt warm.

Q2 List the things which felt cold.

Q3 Which are the best conductors of energy: **A**, **B**, **C** or **D**?

Q4 Which are the best insulators: **A**, **B**, **C** or **D**?

Q5 a Imagine you ride a bike to school. Each morning you notice that the metal handle bars always feel colder than the rubber grips. Can you explain why the two parts of the same bike feel different?
b Why do you think all bikes have rubber grips on the handle bars?

Q6 Write down other examples you have experienced which are similar to **Q5**.

Q7 Explain why cooking pans are made from metal but usually have a wooden or plastic coating on their handles.

4 Storing and transferring energy

Conduction

Conduction takes place in solids. In conduction energy is transferred from atom to atom. The atoms at the source of the energy start **vibrating** (moving backwards and forwards) very fast. These fast moving atoms bump into the atoms next to them which then causes other atoms to start vibrating. In this way the energy is transferred from the source to all the atoms of the solid. Heating by this **process** is called conduction. Heat is the process of making atoms move faster.

Apparatus

- [ ] beaker   [ ] hot water
- [ ] metal teaspoon   [ ] card
- [ ] scissors   [ ] plastic teaspoon
- [ ] gauze   [ ] Bunsen burner
- [ ] tripod   [ ] copper rod
- [ ] iron rod   [ ] aluminium rod
- [ ] brass rod   [ ] glass rod
- [ ] Vaseline   [ ] heatproof mat
- [ ] 5 drawing pins
- [ ] digital stopwatch
- [ ] eye protection

 Wear eye protection when heating materials.

In this experiment you are going to compare how well different materials conduct energy.

Q1 Copy this table.

Time for pin to fall				
Copper(s)	Iron(s)	Glass(s)	Aluminium(s)	Brass(s)

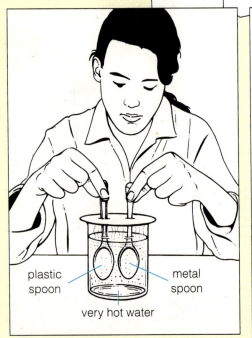

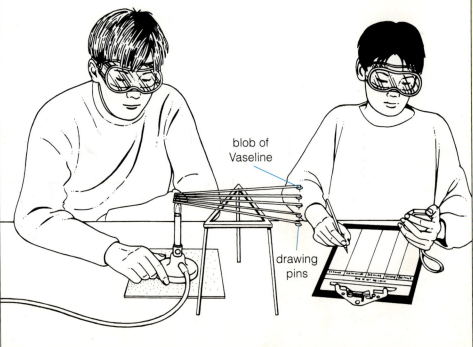

A Push the spoons into two slits in the card. Place them in the hot water. Hold the ends of the spoons. What do you notice after a short time? ▲

B Put some Vaseline on the drawing pins and stick them at the end of each rod. Position them on the tripod as shown and light the Bunsen burner. Note the time taken for the pins to fall from each rod. ▲

Q2 Which spoon gets hot first?

Q3 Which spoon is the best conductor of energy?

Q4 Which way is the energy flowing?

Q5 Did the energy flow at the same rate through all the conductors? Explain your answer.

Q6 Write down the materials in order, starting with the best conductor.

Q7 Which material would make the best saucepan? Explain your answer.

Q8 Why do you think people who stack the large refrigerators in supermarkets are given gloves and thick jackets to wear?

4 Storing and transferring energy

Keeping heat in

The best insulators are materials which trap lots of pockets of air in them. Air is a poor conductor of energy. Work in pairs.

Q1 Copy this table.

Time (m)	No covering (°C)	Wallpaper (°C)	Polystyrene (°C)	Rock wool (°C)
0				
2				
4				
6				

Apparatus

- ☐ 1 uncovered 250 cm³ beaker
- ☐ 3 × 250 cm³ beakers covered with wallpaper, polystyrene and rock wool ☐ 4 beaker lids
- ☐ 4 thermometers ☐ felt-tip pen
- ☐ digital stopwatch ☐ paper
- ☐ 2 stopwatches

no covering   wall paper   polystyrene   rock wool

A Put 200 cm³ of boiling water in each beaker. Put the lid on and push the thermometer through the hole. Record the temperature of the water in each beaker immediately and then every two minutes. Record your results for 30 minutes. ▲

Q2 Which beaker cooled down the quickest?

Q3 Why do we need the uninsulated beaker in the experiment?

Q4 a Which insulated beaker cooled down most quickly?
b What does this tell you about the material?

Q5 a Which insulated beaker cooled down slowly?
b What does this tell you about the material?

Q6 Make a list of the materials starting with the best insulator.

Q7 Where are these materials used in our homes?

Q8 Why are these materials used in our homes?

Q9 Can you explain why during winter snow melts on some roofs before others?

27

4 Storing and transferring energy

Convection

Energy is transferred through a gas or a liquid by **convection currents**. Near the heat source the liquid or gas expands and rises while the cooler (heavier) liquid or gas falls. This keeps happening until all of the liquid or gas is the same temperature. You are going to do some experiments to learn about convection currents.

A At the start the see-saw is balanced. ▼

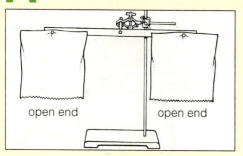

C Tape two balsa wood strips at the opening of the large plastic bag. ▼

E When the bag looks like this let go and watch what happens. ▼

F To show convection currents in water, use forceps to place a small crystal of **potassium permanganate** at the bottom, near the side of a beaker of water. Warm the water with small Bunsen flame. The crystal dye will make the convection current visible. ▶

B Light the candle and place it under one of the bags as shown. Draw a picture of what happens. ▼

D Carefully hold the opening of the bag over the heat source. Make sure it is not too close. Watch what happens to the bag as energy is transferred to the air inside it. ▶

Apparatus

- ☐ 2 light paper bags ☐ pivot
- ☐ large thin plastic bag ☐ beaker
- ☐ toaster *or* shielded heat source
- ☐ metre rule ☐ balsa wood
- ☐ candle ☐ sticky tape
- ☐ Bunsen burner ☐ gauze
- ☐ tripod ☐ heatproof mat
- ☐ potassium permanganate
- ☐ tin lid ☐ eye protection
- ☐ hollow glass tube ☐ forceps

 Wear eye protection.

 Take care not to set fire to paper bags.

Q1 Can you explain what happened in **B**?

Q2 What happens when you remove the candle in **B**?

Q3 Draw a picture to show what happens when you place the candle under the other bag.

Q4 Which is heaviest – hot air or cool air?

Q5 What happens to the bag in **E**?

Q6 Can you explain why this happens?
(*Clue* look back at **B** again.)

4 Storing and transferring energy

Radiation

Infra-red radiation transfers energy from one place to another. When it strikes an object, the energy is absorbed by the atoms of the object. Objects **absorb** (take in) and **emit** (give out) energy all the time. The amount of infra-red radiation absorbed or emitted by an object depends on its temperature and the colour of its **surface**. A special infra-red camera was used to take this photograph. The light blue areas of the face are hottest. Can you work out why? *Hint* look at page 27 again.

Q1 Copy this table.

Time	Temperature of water in shiny can (°C)	Temperature of water in dull can (°C)
Start		
1 minute later		
2 minutes later		
↓		
15 minutes later		

Apparatus

☐ matt black can ☐ shiny can
☐ 2 thermometers
☐ hot water

A Pour equal amounts of hot water into the matt black can and the shiny can. Take their temperature every minute. ▶

B Design and carry out an experiment which will explain why a fire fighter's suit is shiny and not matt black. Write a report showing apparatus diagrams, method and measurements. ▼

Q2 Which can cools down more quickly?

Q3 Which can is emitting energy more quickly?

Q4 Which can is the best radiator of energy?

Q5 Why should the cooling fins on a refrigerator or a motor bike be dull black?

Q6 Why are teapots and kettles bright and shiny?

Q7 Why are houses in hot countries often painted white?

Extension exercise 6 can be used now.

5 Using energy

Saving energy at home

Keeping your house warm can be expensive. The costs can be reduced by better **insulation**. The picture shows two houses: one has no insulation at all, and in the other many steps have been taken to stop the heat from escaping.

Apparatus
- felt-tip pens
- sugar paper
- marker pens

Work in pairs.

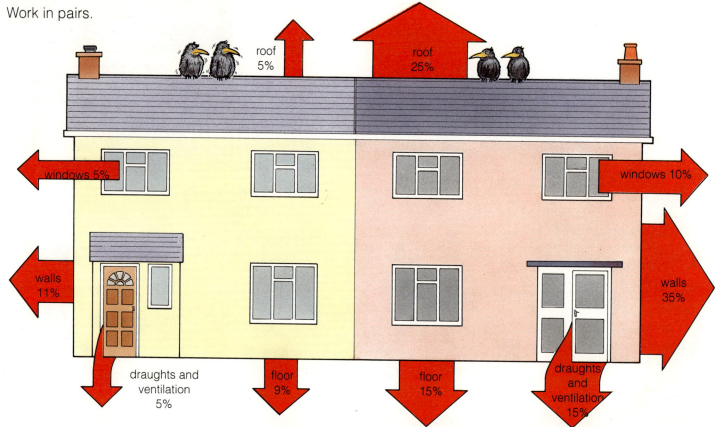

Insulating your house can cost a lot of money. The table shows how quickly improvements pay for themselves.

Improvement	Pays for itself in:
Hot-tank jacket	6 months
Draught proofing	1 year
Roof insulation	1 – 2 years
Wall insulation	2 – 4 years
Room thermostats	2 – 5 years
Double glazing	Over 40 years

There are many other things which you can do to keep warm, such as carpeting floors, using wallpaper, using **thermostats**, putting foil behind radiators, and so on.

Q1 Look at the picture and draw an energy transfer diagram for each house.

Q2 a Where is most of the heat lost from the house with no insulation? Name two places.
b How could the heat loss through these two places be reduced?

Q3 Which house would have the highest heating bills? Explain your answer.

Q4 Which insulation methods pay for themselves quickly?

Q5 Which insulation method takes a long time to pay for itself?

Q6 How do thermostats help to save energy?

Q7 Make an eye-catching poster and help other students to learn how to save energy in your home and school.

Extension exercise 7 can be used now.

5 Using energy

Energy use per person

The big energy spenders are highly industrialised nations like Japan, North America and Western Europe. Gas, oil and coal are used to make electricity, heat homes, work electric gadgets, fuel cars and other transport, power machines in factories, and so on.

The amount of fuel used by people all over the world is different, for example: oil use.

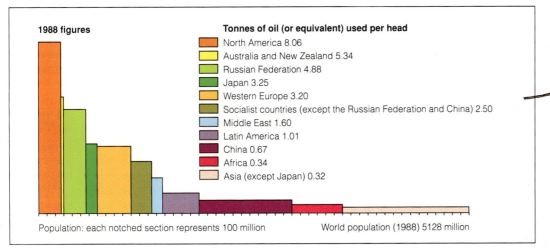

1988 figures

Tonnes of oil (or equivalent) used per head
- North America 8.06
- Australia and New Zealand 5.34
- Russian Federation 4.88
- Japan 3.25
- Western Europe 3.20
- Socialist countries (except the Russian Federation and China) 2.50
- Middle East 1.60
- Latin America 1.01
- China 0.67
- Africa 0.34
- Asia (except Japan) 0.32

Population: each notched section represents 100 million   World population (1988) 5128 million

Q1 How many ways is energy being used in the picture above?

Q2 How much oil (or equivalent) does each of us in Britain use in one year?

Q3 How much more oil does a person in Western Europe use than a person in Asia?

Q4 Which country is the biggest energy spender?

Q5 Which country is the lowest energy spender?

Q6 Give examples of how you use energy every day.

Q7 Give examples of how you think a teenage person in Asia uses energy.

Q8 Look carefully at the graph. What do you notice about the population size and the energy spent per person in the country? Give one reason why these countries are less industrialised.

Extension exercise 8 can be used now.

5 Using energy

World energy problems

No more fossil fuels are being made. We are using up all the gas, oil and coal that there is in the world. Uranium will run out around 2030. When these fuels are gone, there will be no coal left or nuclear power to make any more electricity, no gas to heat our homes and no oil or oil products for our cars or planes. Renewable energy will become very important.

The future

▶ One place that we could investigate is beyond the Earth's **atmosphere** where the Sun's rays are very strong. Orbiting power stations could convert the Sun's energy to electricity using huge **solar cells**. The electricity could be beamed down to Earth using receiving **aerials**.

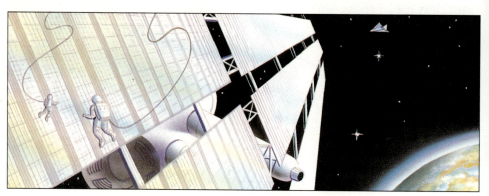

What can we do to save fossil fuels?

▶ **Recycle**

Half the waste from an average home can be recycled. About one third of it comes from packaging. It is made up of glass, paper, metal cans, plastics and food waste. Recycling rubbish reduces the rate at which we use up the Earth's resources. It also saves fuel as less energy is needed to recycle than to process raw materials. Many recycling collection points are in supermarket car parks.

▼ **Use public transport**

If there were fewer cars we would save a lot of oil and live in a cleaner environment.

500 people ⟶ can travel in: ⟶ 125 of these:

or 1 of these: or 10 of these:

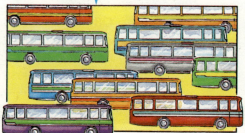

Q1 Work in a small group. Imagine you are a team of scientists about to produce a radio programme on energy. Your aim is to inform the public of the energy problem and what we can do to reduce the use of fossil fuels. Tell them also about the renewable sources of energy available and some of the future plans to capture the Sun's energy.

Q2 When you have planned your documentary record it on a cassette tape and play it to the rest of your class.

LIVING ON PLANET EARTH 200 million years ago would have been a very weird experience. Climate, vegetation and wildlife were all different. Dinosaurs dominated the planet for over 180 million years, but not all creatures on prehistoric earth were dinosaurs. There were many other types of creature in existence, including everything from tiny insects to giant flying lizards.

To be a dinosaur, a creature must:
- have lived from 245 million to 65 million years ago;
- have lived on the land, not in the air or in the water;
- be a reptile (although not all reptiles were dinosaurs);
- have legs located below its body, not protruding from its side (like a crocodile).

Prehistoric lifeforms are often discovered fossilised in rock, but can also be found frozen in ice or trapped in amber. North America has seen some of the most dramatic finds including the recent discovery of a gigantic Tyrannosaurus Rex in South Dakota. Named Sue after its finder, it is the largest and most complete Tyrannosaurus anywhere in the world – so far!

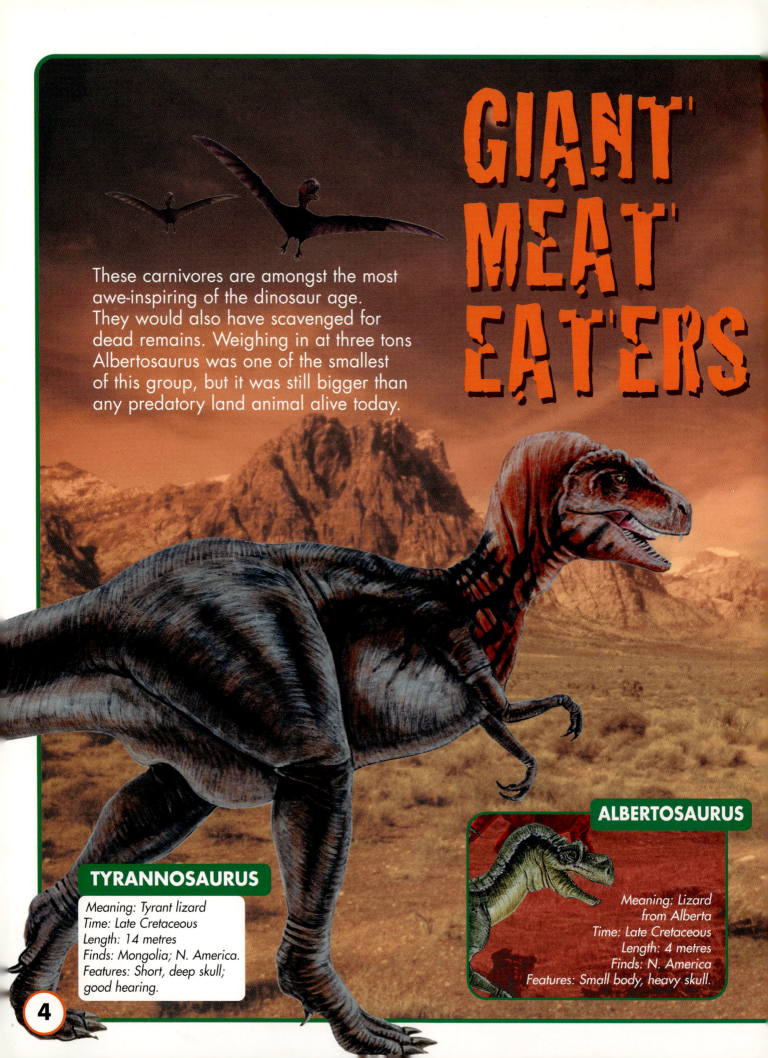

GIANT MEAT EATERS

These carnivores are amongst the most awe-inspiring of the dinosaur age. They would also have scavenged for dead remains. Weighing in at three tons Albertosaurus was one of the smallest of this group, but it was still bigger than any predatory land animal alive today.

TYRANNOSAURUS

Meaning: Tyrant lizard
Time: Late Cretaceous
Length: 14 metres
Finds: Mongolia; N. America.
Features: Short, deep skull; good hearing.

ALBERTOSAURUS

Meaning: Lizard from Alberta
Time: Late Cretaceous
Length: 4 metres
Finds: N. America
Features: Small body, heavy skull.

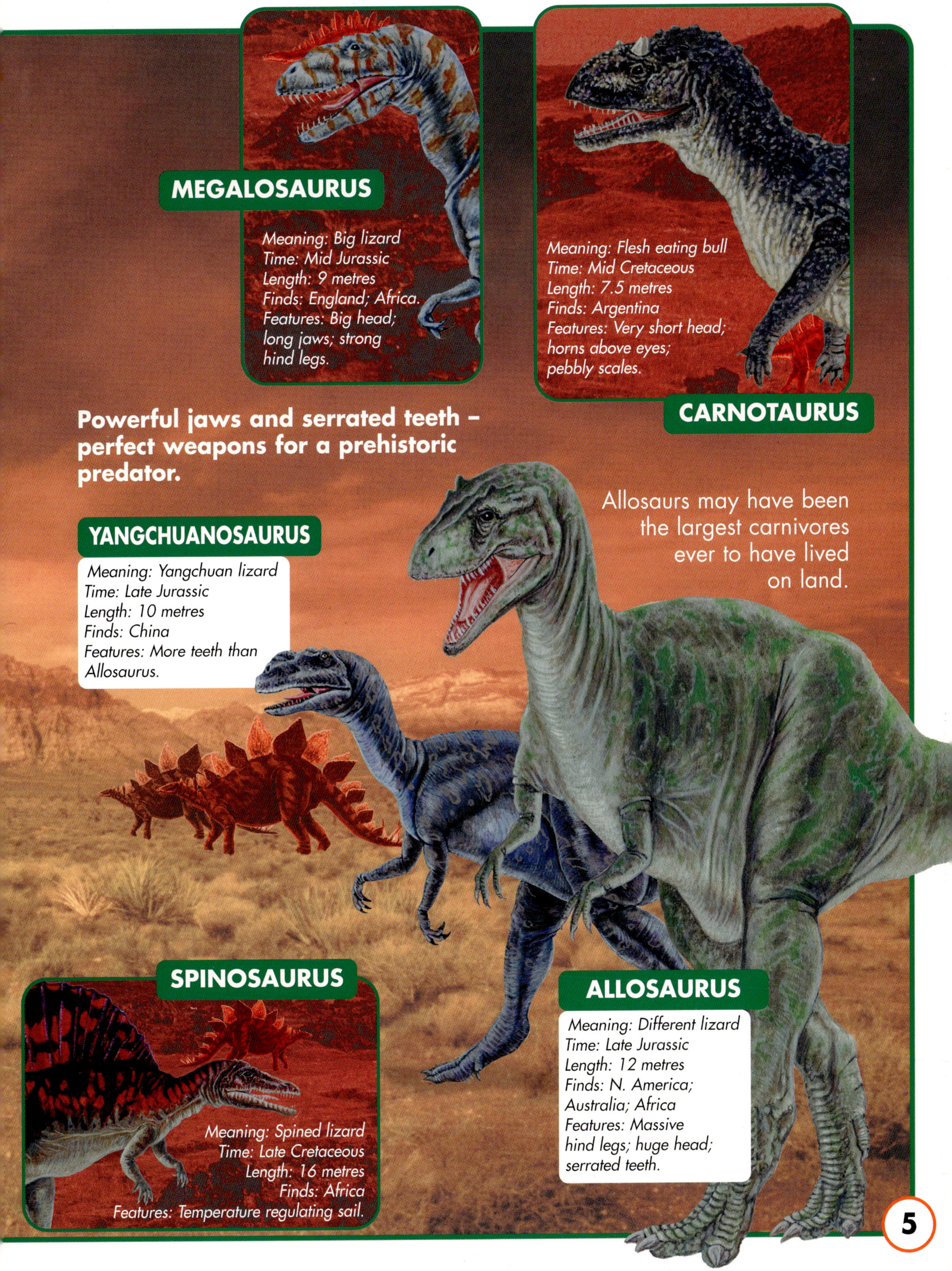

MEGALOSAURUS

Meaning: Big lizard
Time: Mid Jurassic
Length: 9 metres
Finds: England; Africa.
Features: Big head; long jaws; strong hind legs.

CARNOTAURUS

Meaning: Flesh eating bull
Time: Mid Cretaceous
Length: 7.5 metres
Finds: Argentina
Features: Very short head; horns above eyes; pebbly scales.

Powerful jaws and serrated teeth – perfect weapons for a prehistoric predator.

YANGCHUANOSAURUS

Meaning: Yangchuan lizard
Time: Late Jurassic
Length: 10 metres
Finds: China
Features: More teeth than Allosaurus.

Allosaurs may have been the largest carnivores ever to have lived on land.

SPINOSAURUS

Meaning: Spined lizard
Time: Late Cretaceous
Length: 16 metres
Finds: Africa
Features: Temperature regulating sail.

ALLOSAURUS

Meaning: Different lizard
Time: Late Jurassic
Length: 12 metres
Finds: N. America; Australia; Africa
Features: Massive hind legs; huge head; serrated teeth.

5

Ceratosaurus was a tyrannosaur. It had the same tiny, feeble arms as other tyrannosaurs.

> Ceratosaurus were related to the Tyrannosaurus Rex, which was the king of the meat eaters. Ceratosaurus grew to 6m (20ft) in length.

HEAVY METAL

This group of dinosaurs are ancestors to the modern-day rhinoceros. The aggressive looking headgear was actually for defensive purposes. Distinctive heat regulating plates on its back make Stegosaurus one of the most recognisable of all dinosaurs. Stegosaurus evolved in China before migrating to Africa and North America.

TRICERATOPS

Meaning: Three-horned face
Length: 9 metres
Finds: N. America
Features: three skull horns; large neck shield.

Many plant-eating dinosaurs may have swallowed stones to help digestion.

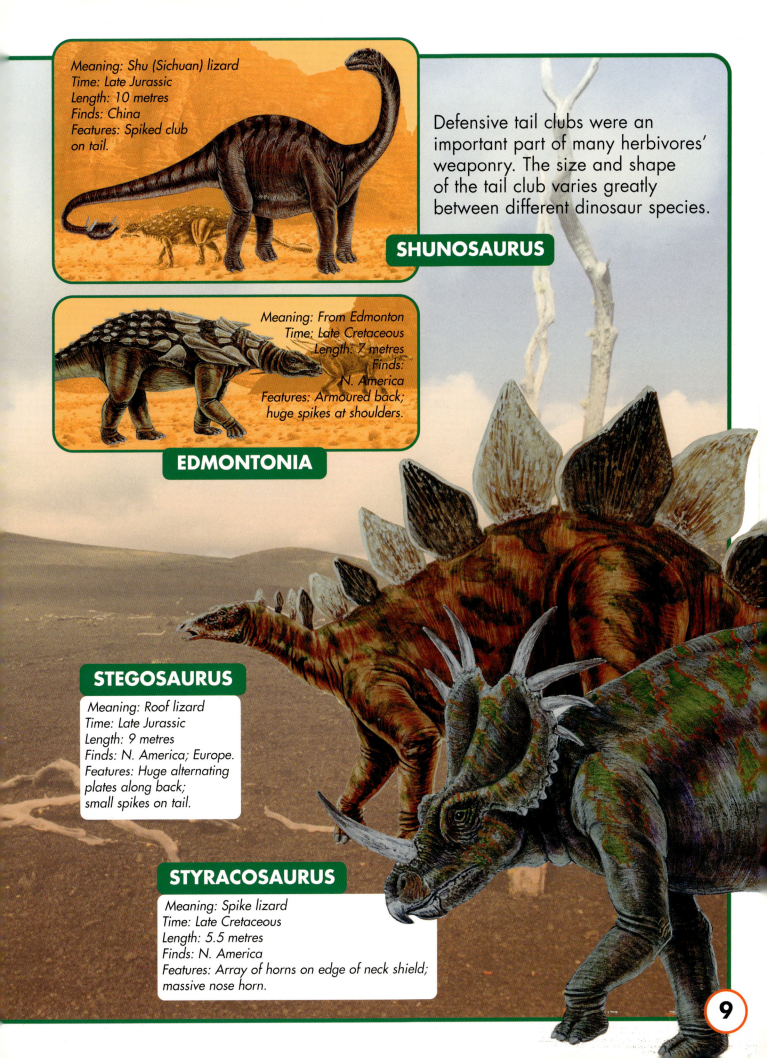

Meaning: Shu (Sichuan) lizard
Time: Late Jurassic
Length: 10 metres
Finds: China
Features: Spiked club on tail.

Defensive tail clubs were an important part of many herbivores' weaponry. The size and shape of the tail club varies greatly between different dinosaur species.

SHUNOSAURUS

Meaning: From Edmonton
Time: Late Cretaceous
Length: 7 metres
Finds: N. America
Features: Armoured back; huge spikes at shoulders.

EDMONTONIA

STEGOSAURUS

Meaning: Roof lizard
Time: Late Jurassic
Length: 9 metres
Finds: N. America; Europe.
Features: Huge alternating plates along back; small spikes on tail.

STYRACOSAURUS

Meaning: Spike lizard
Time: Late Cretaceous
Length: 5.5 metres
Finds: N. America
Features: Array of horns on edge of neck shield; massive nose horn.

The 7m (24ft) Stegosaurus was a plant-eater that probably lived in herds.

Stegosaurus lived 145-200 million years ago. They weighed around 1.75 tonnes and only ate soft plants.

open these pages to see your...

DINO

FAST ATTACKERS

COELURUS
Meaning: Hollow tail
Time: Late Jurassic
Length: 2 metres
Finds: NAmerica
Features: Sharp, curved teeth.

COELOPHYSIS
Meaning: Hollow form
Time: Late Triassic
Length: 3 metres
Features: Lightweight running hunter; strong skull and neck.

Unusually large brains were a feature of this group. Large eyes indicate they may have been well suited to hunting at night.

VELOCIRAPTOR
Meaning: Fast hunter
Time: Late Cretaceous
Length: 4 metres
Finds: Mongolia; China.
Features: Eagle-like talons; curved killing claw.

18

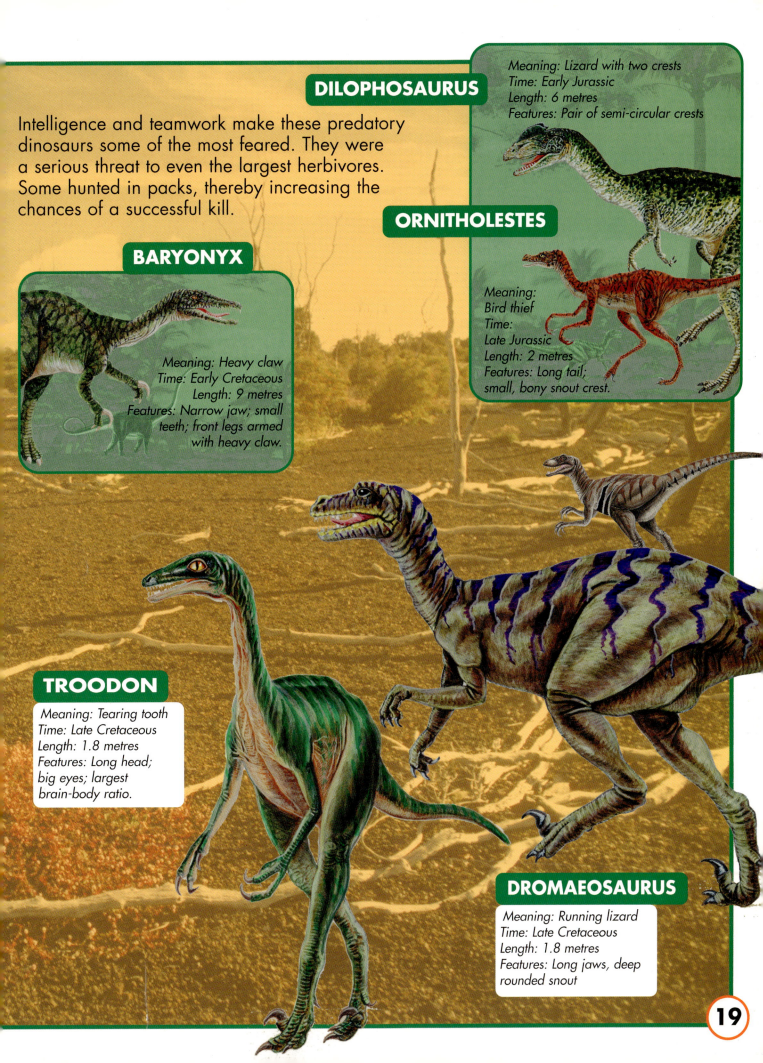

Intelligence and teamwork make these predatory dinosaurs some of the most feared. They were a serious threat to even the largest herbivores. Some hunted in packs, thereby increasing the chances of a successful kill.

DILOPHOSAURUS
Meaning: Lizard with two crests
Time: Early Jurassic
Length: 6 metres
Features: Pair of semi-circular crests

ORNITHOLESTES
Meaning: Bird thief
Time: Late Jurassic
Length: 2 metres
Features: Long tail; small, bony snout crest.

BARYONYX
Meaning: Heavy claw
Time: Early Cretaceous
Length: 9 metres
Features: Narrow jaw; small teeth; front legs armed with heavy claw.

TROODON
Meaning: Tearing tooth
Time: Late Cretaceous
Length: 1.8 metres
Features: Long head; big eyes; largest brain-body ratio.

DROMAEOSAURUS
Meaning: Running lizard
Time: Late Cretaceous
Length: 1.8 metres
Features: Long jaws, deep rounded snout

It stood less than 1m (3ft) high, but Deinonychus, "terrible claw" could rip open a bigger dinosaur!

Deinonychus was a fast, agile and aggressive meat-eating hunter. It used razor-sharp teeth and claws to kill its prey.

SUPER LONG NECKS

With their heads so far from their bodies, how did these huge dinosaurs not faint? One theory is that they may have had more than one heart.

A real giant of the prehistoric world, Brachiosaurus is one of the most famous dinosaurs ever. Long front legs helped it reach twice the height of a giraffe.

BAROSAURUS

Meaning: Slow heavy lizard
Time: Late Jurassic
Length: 27 metres
Finds: N. America; Africa.
Features: Extremely long neck.

BRACHIOSAURUS

Meaning: Tall-chested arm lizard
Time: Late Jurassic
Length: 26 metres
Finds: N. America; Africa; Europe.
Features: Long neck and front legs.

DICRAEOSAURUS

Meaning: Two forked lizard
Time: Late Jurassic
Length: 14 metres long
Finds: Africa
Features: Shorter neck than most diplodocids.

MAMENCHISAURUS

Meaning: Lizard from Mamen Brook
Time: Late Jurassic
Length: 25 metres
Finds: China
Features: The longest neck of any known dinosaur – 14m.

MUSSAURUS

Meaning: Mouse lizard
Time: Late Triassic
Length: 3 metres
Finds: Argentina
Features: Small head; long tail.

ASTRODON

Time: Early Cretaceous
Length: 18 metres
Finds: N. America
Features: Similar to Brachiosaurus.

Long necks are the perfect tool for tree grazing. Some long necked dinosaurs could graze large areas without needing to move.

The first dinosaur bone was found in 1676. About 150 years later, in 1824, the bone was described as part of a giant fossil lizard.

Even today, only a tiny number of the bones left behind by dinosaurs have been found. So one day, you could find a dinosaur bone!

WINGED WONDERS

THE FIRST FLYING REPTILES appeared over 240 million years ago, simply gliding between trees. Time and evolution eventually brought the Pterosaurs, which took to the air on wings of skin. They ruled the skies while dinosaurs ruled the land. Some Pterodactyls were no larger than pigeons, while others were the largest flying creatures ever to have existed.

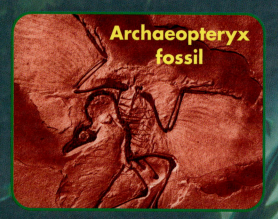

Archaeopteryx fossil

Pterosaur fossil

The first evidence of feathered prehistoric creatures was discovered in Germany in 1861. This fantastic fossil of Archaeopteryx clearly shows its feathered wings and tail – although it did also have some reptilian features – a toothed beak for example.